I0463465

An Intro to Acupuncture And TCM
(Traditional Chinese Medicine):

How To Lose Weight, Feel Great, And Fix Your Sore Back
With Acupuncture And Other Techniques From
Integrative Health Care In China

Martin Avery

Dedication: To Dr. Guan "Sophia" Wang, the best doctor I've ever met on the planet.

ISBN 978-1-304-92063-8

Copyright © Martin Avery, Dalian, China, 2014

PS: A special "thank you" goes to Dr. Wang for reading a draft of this book.

An Intro to Acupuncture TCM
(Traditional Chinese Medicine):

How To Lose Weight, Feel Great, And Fix Your Sore Back
With Acupuncture And Other Techniques From
Integrative Health Care In China

Martin Avery

A book about Acupuncture is a super, super idea. Everybody knows approximately two things about it, and everybody would like to understand it better. Should, too.
- Canadian publishing legend, Anne Millyard

Nearly 8 out of every 10 people will have low back pain sometime in their lives. It is one of the top reasons people seek medical treatment. It is the number one reported reason for trying acupuncture. The good news is that low back pain is one of the many conditions acupuncture may be an effective tool for treating. - WebMD Medical Reference

Traditional Chinese Medicine relies on thousands of years of experience; its validity and effectiveness have been documented and reported by authoritative sources including the World Health Organization.

Introduction

In 2014, I had a great opportunity to explore Traditional Chinese Medicine in China with a highly trained, talented, and gifted doctor. Over the period of a month, I had acupuncture every other day in various combinations with moxibustion, heat, finger needling, fire cupping, herbs, and qigong, not to mention nutrition information and the philosophy behind it all.

One of my goals, for going to China, was to learn a few things about Traditional Chinese Medicine. I come from Norman Bethune's hometown and I had a desire to do the opposite of what Bethune became so famous for. He took Western medicine and surgical techniques to China. I wanted to take home some Traditional Chinese Medicine back to Canada.

Before going to China, I experienced acupuncture in Ontario and Alberta, in Canada, and had positive experiences, but I have to say .... Acupuncture as a part of TCM, rather than by itself, is far superior.

Another key element is the doctor who does the work. I was very fortunate to be introduced to Dr. Guan "Sofia" Wang. She insists that the doctor and the patient work together, that it takes both people doing the right thing, but .... I don't know! I think she deserves all the credit!

In order to write this book, I kept a journal of my visits to the doctor's office. To give you a feel for the process, I decided to structure the book using those journal entries. Acupuncture is not something you do once, I discovered. For the best results, many sessions are required.

To make a long story short, I dropped a size and fixed my back!

Acupuncture is a component of the health care system of China that can be traced back at least 2,500 years. The general theory of acupuncture is based on the premise that there are patterns of energy flow (Qi) through the body that are essential for health. Disruptions of this flow are believed to be responsible for disease.

Now more than ever before, people across North America are returning to more natural methods of improving their health. Traditional Chinese Medicine (TCM) is continually growing in acceptance and practice in Canada and the U.S.A.. It focuses on health promotion, illness prevention, and treatment through natural remedies such as acupuncture, Chinese herbal medicine, and tuina massage. The philosophy and practice of TCM are consistent with the most significant developments in healthcare today:

1. An emphasis on prevention.
2. The call for natural, non-invasive medical approaches.
3. An acknowledgement of the need to balance the physical, emotional and spiritual aspects of life.
4. The changing role of the medical practitioner as the facilitator of a patient's health and well-being.

The study and practice of TCM brings together traditional and modern views of health and eastern and western approaches. It provides an opportunity to participate in one of the truly remarkable achievements of mankind. TCM has served over a quarter of the world's population for thousands of years and repeated clinical tests have proven its effectiveness in fighting many diseases.

In China, its country of origin, TCM and western medicine work hand-in-hand to provide the very best health care and disease treatment possible.

Across North America two trends are emerging in this exciting field: 1. the recognition of TCM as a formal health care profession after only two thousand years and 2. integrative health care, combining the best of the East and the West.

My Back Story

Last year, when I was in Cold Lake, Alberta, Canada, I went to a chiropractor, for the first time.

I had some back pain and could not get a doctor and the hospital didn't have a walk-in clinic.

So, I went to a chiropractor.

The Wellness Centre had a waiting period, so I went to one down the road. He adjusted bones, took away back pain, but after a few days, I could no longer walk! My legs were numb! The chiropractor said, That happens, sometimes; you should go to 'emergency' at the hospital and get some pain killers.

A Cold Lake doctor in the emergency unit at the hospital sent me to an emergency neurology specialist in Edmonton.

He said, Activate your short and long term disability because you will never work or walk again.

He said, You have GBS, Guillaine Barre Syndrome, and you're going to die.

Lucky for me, I had read Joseph Heller's book, No Laughing Matter, about the time he had GBS, and I knew the doctor was wrong, so I laughed, got up, and walked out! -- The chiropractor hit my sciatic nerve bundle, causing nerve damage, that shut down the muscles from my knees to my toes.

Before I went to China, I tried acupuncture in Canada. I tried "Western" acupuncture with Dr. Sabrina Hooper at Your Health Chiropractic in Port Perry, Ontario, to fix plantar fasciitis. I found a great reflexologist in Cold Lake, Alberta, named Mecell Pilon, and a physiotherapist at Health & Sport Rehab, also in Cold Lake, who also did acupuncture.

Mecell Pilon, a member of the Cold Lake Chiropractic & Wellness team, is an Alternative Practitioner, a Certified Instructor of Touchpoint Reflexology, known for pain management, is also a Reiki Master and Applied Kinesiologist and, I would say, a very gifted healer.

Dr. Wang, my TCM doctor, uses acupuncture, finger needling, or tuina, moxibustion, heat lamps, and herbs, plus nutrition, to rebuild the muscles around my spine and to regenerate the peripheral nerves down to my toes. She says nerve regeneration doesn't happen very often, but it's happening for me!

1. The First Visit

The first time I went to see Dr. Wang, I was very curious. I had heard a lot of good things about her from my colleagues. The woman who introduced me to her (let's call her 'Beth') said, "She got me through last year, when I had a variety of health concerns, from carpel tunnel syndrome to coughs and colds and so on."

The clinic was as big as the general hospital in my old hometown, or in any town in Canada. It was open to the public and attached to lots of other clinics, which offered a full range of services, without waiting, including x-rays, CT scans, MRIs, and also dental services.

In Ontario, the last time I moved, I had to wait for six months, before a doctor would add me to her huge list of patients. In Alberta, there wasn't even a waiting list. All the doctors came from South Africa and none of them were taking new patients. I had to get a doctor in the next town, an hour and a half away, by car, so every visit to the doctor meant a three hour round trip.

Dr. Wang was waiting for us, with no patients in a waiting room to worry about, and she gave us all the time we wanted. And she did not deal with just one ailment, the way doctors do in Canada. First she talked to Beth about whatever she wanted to talk about. Dr. Wang gave her acupuncture for a sore wrist the year before and wanted to know how she was doing.

She struck me as being very compassionate and a very good listener, unlike any Canadian doctor I had ever encountered. She had a look of concern on her face and a consoling tone in her voice. She made eye contact and nodded her head, smiled and frowned, when appropriate, showed she had a sense of humour but maintained a professional air, while being friendly.

Perfect, I said to myself.

When it was my turn, she listened intently, asked good questions, and made me feel heard, or listened to, more than any doctor I had ever met anywhere.

I told her my hamstrings hurt, sometimes, so she asked a lot of questions so she would know when it hurt and how much. She got me to twist and turn and bend and squat, so she could find out what I could do before I felt any pain.

Anything else? she said. I told her my ears get blocked sometimes. She asked me all about that, too.

After twenty minutes of discussion, she invited me to take off my shirt and pants and climb onto her table.

She also looked at my tongue, but did not say anything about what she noticed.

She had two massage tables set up so two people could get treated together.

While she was getting her long, thin, needles ready, I said something about a poster she had on her wall, which showed the meridians and points on the body that acupuncturists used. I told Beth I'd like to get one and the doctor responded by pulling one out of a drawer and handing it to me.

She made sure I was comfortable on the massage table and then looked after my ears and my hamstrings. She dropped cold ear wash into one ear and then the other and massaged the area around the ears and then popped a row of three needles into the backs of both my legs. She put a heat lamp over my legs and asked me if it was too hot or uncomfortable.

When I was dressed and ready to go, she gave me her bill, for just over 1000 Rmb, or just over 160 bucks, Canadian dollars.

In addition to the payment, I gave the doctor a little present.

Before I left Canada, I visited the Bethune House, in Gravenhurst, my old hometown, to tell them I was going to China and to see if they had any suggestions or advice. Bethune, as most Canadians and approximately 1.5 billion Chinese people know, was a doctor who left Canada for China to work as a surgeon on the battlefront in their War of Resistance against Japan.

The Bethune House, where Norman Bethune was born, is now a museum.

They told me I should take some of their buttons with me, when I went to China, explaining that it is a small but important part of Chinese culture to exchange gifts.

They had run out of buttons, so I had my own buttons made up, using a picture I took of a life-like sculpture of Bethune in front of the Bethune House.

The doctor was so touched by the little present, she told me how she felt about the Canadian doctor.

She said, He gave his life to the cause so Mao wrote about him in his famous "Little Red Book" and he became a hero to generations of Chinese people. Mao praised him for his selfless devotion. Many also give Bethune credit for introducing Western medicine and surgical techniques to China, especially the rural areas where he worked with the Eighth Route Army.

That was has been described as the greatest catastrophe of human history, unexampled in the destitution of millions. The war of the Chinese people against the Japanese invaders was the fight of one-fifth of the human race for national independence -- for life, liberty and the pursuit of happiness. It was also a war in which hungry, half-armed Chinese farmers held the front lines against imperialism which threatened Asia and the peace of the world. It is also seen as the time the Chinese people began to awake, to unite, to become a nation.

Bethune took his Mobile Army Support Hospital concept to China and travelled with the Chinese Eighth Route Army and is given credit for saving thousands of lives. He also taught the people he worked with to be doctors, nurses, and orderlies. And he died with his boots on, operating in a former temple in the mountains, not far from the battlefront.

Now there are hospitals, awards, and honours named after Bethune, in China, and there have been books, movies, and a high profile TV series about Bethune in China.

I was already well aware of Bethune's history but it felt very gratifying to hear how he was regarded by a young doctor in China.

She said, Bethune still has a big place in the hearts of Chinese people.

When I walked out of the clinic, I felt pretty good, and thought I was grounded, or in the game, but later on, looking back, I realized I was feeling a little strange. I felt as though I was floating a foot or so off the ground. It was a very good thing I had someone with me, especially somebody who was good at negotiating with taxi drivers and knew the route back to downtown Kaifaqu. I felt as though I was still out of my body, after lying in the heat lamps and getting acupuncture.

I could walk! I could hear! It was a miracle! -- That's the way it felt.

The doc gave me ear drops and some herbal capsules to take over the next few days and said it would be best if I could return for more acupuncture a few times in the coming week but understood that teachers were very busy.

Come back when you can, she said, or when you want to.

I thanked her and said, See you later!

2. The Second Visit

My second visit to the TCM doctor was just like the first, for me, but it was quite different for my friend, who was treated on the massage table next to mine.

The doctor was wearing her Bethune button on her lab coat.

For the second time, I had several needles in each leg, a moxibustion box on top of each leg, and heat lamps.

It felt as though I was on the table for a long time, but it was only twenty minutes. During that dreamy time, I thought about the novels and other books I was writing, and just generally enjoyed the warmth of the heat lamps.

I had had a cough or a cold for the first month after I landed in China and it was finally fading away.

Lying on my stomach, with my face in the hole in the headrest, designed for that purpose, your nose runs, if you have a cold or the flu.

I got up from the session feeling much better in every way, just as I had the first time.

My friend gave the doctor a list of complaints, including the fact that she didn't want to be there, and I had dragged her there. She said she had a sore wrist, a sore shoulder, headaches, a cough, she wanted to lose weight, and there might have been a few other things. I lost track, but the doctor didn't.

The doctor got her on the massage table, lying on her back, and gave her a lot of needles for her wrist and shoulder, one in the top of the head, for headaches and body pain, and several in her stomach area, for weight loss.

When her acupuncture session was over, she had a second session, as the doctor prepared a seed patch for her ear and attached it for her while explaining what it was all about.

The skin-coloured, sticky, patch, like a bandaid, contained a few seeds. She placed it strategically so each seed was on a specific point. And the doctor told her how to use it, how often to use it and what it was for.

Each seed was on an acupuncture point related to weight loss. More specifically, they influenced appetite. A decreased appetite would lead to weight loss, she said.

She had been her doctor for over a year and already knew what she ate, when she ate, how much she ate, and so on.

The patch with seeds was invisible, in her ear.

A lot of people use them, the doctor said. You probably never notice them, but if you look closely or carefully at the ears of people on the train or the bus or wherever you are close to them, you will see a lot of people wearing patches on their ears. And you might see them touching their ears. They are massaging the points, doing their own acupuncture, for different things.

The ear patches and ear acupressure or acupuncture was a Chinese invention, she said. Acupuncture of the ear has been popular in China for centuries. However, what has been known as auricular therapy was developed in the 1950s when a French doctor named Paul Nogier, mapped specific sites, or points, on the ear based on the shape of a fetus (or homunculus).

Take a careful look at a human ear, she said, probably someone else's, and see if you can detect the outline of an inverted fetus. This may challenge your imagination, but the concept is important -- it is the key to understanding how auricular therapy works. This "ear map" is how the acupuncturist finds the particular points that correspond to the problem area of the patient's body.

She told us she likes to use auricular therapy in conjunction with other types of acupuncture, though many practitioners treat certain types of problems with ear therapy alone.

It is good for many things, including weight loss, to cut back on or stop drinking alcohol, and to stop smoking.

She says she secures about five to seven seeds per ear under a small adhesive dressing (like tape), which should stay in place until the next treatment. Patients learn how to massage the points where the seeds are, thereby activating their effect at least three times a day and/or when they need relief from pain or anxiety or have an urge for the forbidden substance. This massaging may even help them when the seeds are removed.

She said some of her patients stopped a bad smoking habit after just one session, but that the general routine for an addiction treatment is six weekly sessions and then perhaps continuing another month or two after those.

Some patients need "tune-up" therapy from time to time, and she noted that patients who seek auricular therapy for chronic pain or depression may want to return for a treatment every few months or so to keep their energy flowing freely.

If you're suffering from addiction, cravings or unwanted habits, acupressure using ear seeds might be for you. In this form of traditional Chinese medicine, or TCM, a practitioner places the tiny seeds of the vaccaria plant on key points on the ear and tapes them in place. Pressing on the seeds stimulates the related point and might be able to help you reduce cravings when they arise.

According to TCM, the ear hosts acupressure points that correspond to each part of the body. When you stimulate a point, the smooth and abundant flow of qi or vital life energy returns the related organ or area, allowing healing to take place. Ear seeds can be used alone or in combination with acupuncture or body acupressure in order to treat a host of symptoms, and can offer gentle support for your weight loss or smoking cessation efforts, or to help you fight addictions.

The fee for my second visit was lower than the first. It was 450 yuan, or about $75.00 CDN. It felt like the biggest bargain in the world.

3. Appointment #3

When you are new to China and don't speak Chinese, it is a bit of a challenge to get around, so getting to the Dalian Free Trade Zone Hospital for my third appointment, and my first solo visit, taking a taxi to the train station, the train to KaiFaQu, and a taxi to Greentown, felt like an accomplishment.

"You have to count all the small victories. I do, anyway," the American poet Richard Brautigan said.

The appointment went well, again, as the doctor used the same techniques, which I had enjoyed the previous times. She did acupuncture with heat lamps and moxibustion.

At one point, the moxibustion box got to be too hot and I jumped, a little, involuntarily, but did not set the carpeted room on fire.

"Sorry!" the doctor said.

She made sure I was okay and we carried on.

The third time I had acupuncture, I felt the time passed much more quickly. And when it was over, I did not feel as sleepy as I did the first time. I felt energized.

4. Visit #4

My 4th trip to see Dr. Wang was the best yet.

It was a little challenge getting to the clinic as I discovered there were no taxis where I live, at this time of year, so I thought I would be late. I texted the doctor and she texted me back, saying, "Don't worry." We texted back and forth a few times as I got a ride to the Jinshitan train station with a colleague, caught the train on time, and got a taxi right away at Kaifaqu Station. So, I made it in time for the appointment.

The doctor congratulated me for making it on time as she knew there were no taxis where I lived, during the Spring Festival, but she told me not to worry or hurry. "Just relax," she said. "If you come at eleven o'clock instead of ten thirty, it's no problem."

We talked for fifteen or twenty minutes before I got on the table for acupuncture.

I told her that my feet definitely feel different. "They are more sensitive," I told her. "This morning I stepped on something very small and thin, at my place, and actually noticed it. A little while ago, I could stand on something much bigger without being aware of it. So my feet are getting more sensitive."

"That's great," she said.

"Also," I told her, "yesterday I took the train to downtown Dalian and walked around for about three hours. It was chilly but I had a good time exploring the Russian Street, Tianjin; Friendship Square and Victory Plaza; as well as YouHau Street, from end to end."

"You did all that on your own?" she said. "You are a hero!"

She said, "I need someone to show me around Dalian because it is so big and it is so easy to get lost."

She also said, "You walked for three hours in the cold? And your back did not hurt?" She said, "If I walked for one hour like that, my legs and my back would hurt. You must be getting better!"

I told her I loved Victory Plaza.

"You didn't get lost?" she said. "It's so crowded and there are so many little shops on five different floors, all underground. I always get lost. If I buy something from a store, I can never find the store again when I go back. There are very few signs giving directions, and they are all in Chinese. How did you find your way around? I lose my sense of direction when I am underground."

I told her that I loved it, all of Dalian, especially Victory Plaza, but my favourite was the street food at all the booths on both sides of the road leading to the tunnel under the train station. "Great food!" I said.

"You like that?" she said. "You like Dalian? And you explored all on your own? That is very impressive!"

She said she really liked Dalian, too, even more than Beijing, which she loved.

"Better air, because we are right beside the sea," she said.

"Dalian is like San Francisco," she said. "The hills, the city, the water ...."

She worked in San Francisco for a number of years, she informed me.

I told her I wanted to write a travel book about Dalian and a novel set in Dalian and I had just finished writing a book about the Dalian Ice Dragons Hockey Team.

"You should meet the mayor," she said. "He would be very happy to hear that a Westerner loves Dalian and writes books about Dalian. In the U.S.A. and Canada, most people know only the biggest cities of China, and they think it is still like the early 1900s or the 1950s. The mayor would like the world to know what kind of a city Dalian is now."

"Yes," I said. "Everybody has heard of Beijing, Shanghai, and Hong Kong, but a lot of people think the cities of China are still full of opium dens and the streets are so crowded with bicycles that you can't get across the road to get to the opium den."

"Bicycles and motorcycles are no longer allowed in downtown Dalian," she said, "but you still see them in Kaifaqu and Jinshitan."

"I know!" I said. "And I couldn't find any opium dens, either."

That made her laugh.

I told her my favourite street food was a pancake made with an egg, fried in front of you, with your choice of meat added, plus lettuce, two types of sauce painted on with a paint brush, plus some other choices.

"Yes!" she said. "That's my favourite, too!"

She smiles and gets excited when she talks about pancakes.

"When I worked in Beijing," she added, "there was a place that made those pancakes right in front of the place where I worked, and I loved going there every day, from breakfast or lunch."

She told me about the province on the other side of the big bay from our province and how they made a very good pancake there that was quite different. "Thinner," she said, "and a little salty, with meet, but still very good."

"Ah," I said. "Isn't that Mo Yan's province? The guy who won the Nobel Prize in Literature."

"Yes," she said. "China's first winner for the Nobel Prize in Literature."

"Yes," I said. "I've read all his books, in translation, of course."

"You have?" she said. "That's very good!"

She grabbed my shoulder to congratulate me for taking such a big interest in China and its literature."

I told her that the year after China finally won its first Nobel Prize in Literature, Canada finally won its first Nobel Prize in Literature.

"Alice Munro," I said. "I met her a few times and I taught her niece and nephew."

"You know such a famous person?!" she exclaimed.

"I think you will be like Mo Yan and Alice Munro and you will win the Nobel Prize for Literature, for your 100 books in the West and 100 books in the East," she said.

"We need to get someone to nominate you," she added. "I hear that is a key part of the process."

She got me on the table, with my pants off, and needled me in some new places. She put two needles in my lower back, one on each side of the last vertebrae, and she placed several needles in both legs, where she had placed them on previous visits.

She checked my knees, on the insides of my legs, to see if some key spots were sensitive or not.

"Not so much," I said.

"That's good," she said. "If it's painful, it means the energy is blocked, so it's not moving."

After she pops all the needles in, always checking to see if it was painful, she adjusts each one, to make sure I can feel the buzz or tingle like electricity.

That's the way she described it.

"Can you feel it moving down your leg?" she said.

"I don't think so," I said. "Let me focus."

Although I had never felt anything going from one pin to the next, I tried to focus my awareness, to discover if I had that kind of a feeling in my legs.

After a few minutes, I thought I could feel a little electric current going down a short section of both my legs.

"That's good," she said. "You will feel more like that as your back and legs get better."

After needling me, and adjusting the needles, she put a heat lamp on, over my legs, and the she placed the little boxes with burning herbs on top of my legs. She put a little blanket over my feet and ankles.

As usual, she left me, to go to her desk and do some work, but checked in on me regularly, to make sure nothing hurt or got too hot.

At one point, the herb box did get a little too hot, so she moved it away, quickly.

On my earlier visits, the time I spent on the table went slowly. It felt as though I was there for an hour. I had time to think about many things, come up with novel ideas, and so on. But this time, it felt as though it was over after a few minutes.

When I told her, she said, "That's good!"

She put two cups on my back, again, like the last few times. First she pricked me with a needle and then she put a cup over the spot where she had pricked me.

"Sorry, sorry," she said.

"It didn't hurt," I told her.

"Good," she said.

After cupping and needling, she told me I could get up, but coached me to move slowly.

She brought me a box of cookies and urged me to take one. "Chocolate chip with red beans," she said. "Moist, not dry."

I thanked her and tried one, then told her it tasted great.

Next on the agenda, after I got dressed, except for my shoes, was the hand-holding and walking exercise we had done before. She took my right hand in her hands and felt around with her thumbs until she found a sensitive spot by the upper knuckle of my middle finger. When I said, "That's a little painful," she said, "Oh," and pressed harder.

She got me to walk forward while she walked backwards, up and down her office, and around the tables, encouraging me to walk tall, and asking me about any pain.

After a few minutes, I got a little pain in my left leg, at the back. "It feels tight and it's just a little painful," I said.

"Oh," she said, frowning.

She got me to walk tall while she pressed the sensitive spot on my hand some more.

After I sat down, to get ready to go, the pain, or tightness, faded away.

"I feel energized," I said. "I feel like dancing. -- Or something!"

"That's good," she said. "You have a lot of energy. You are very strong and you can walk for hours and hours."

I asked her about qigong in China. First I told her I had been asked to teach Zen meditation alongside a yoga teacher. It was the yoga teacher's idea. We called it Zen yoga.

"Oh, you teach meditation!" she said. "Very good!"

I told her that the Zen master who taught me about meditation also did energy exercises.

"He was a Buddhist, so he didn't want to say the energy exercises were qigung," I told her, "but I had already studied qigung and it looked like the same thing. But, being a Buddhist, he didn't want to have anything to do with qigung as it was related to kung fu and Chinese martial arts, which are competitive and violent, and not Buddhist, so he called it energy exercises."

"Ah, yes," she said. "In Chinese history, Confucianism, Taoism, and Buddhism, go together, and cannot really be separated, and they go together with the roots of Traditional Chinese Medicine."

I asked her about nutrition.

"What kind of food should I eat, to help the healing?" I asked her.

"Back home, I was told that onions would be good, to decrease inflammation," I added.

"Onions are good for colds and for inflammation caused by a virus or bacteria," she said. "But your inflammation is caused by an injury, so it's different."

I asked about food again.

"You might not like what I suggest, this time," she said.

"Oh?" I said.

"Gou rou," she said. "Many people in your country and the rest of the West do not like the idea of eating this."

I had a feeling that I knew what she was talking about.

"It's the meat of the dog," she said. "Chinese restaurants do not serve it or sell it," she added, "but Korean restaurants in China can cook it for you. Some of them."

"Some of the Canadians I work with have tried it," I said. "They say it's no big deal."

"Well, you can try just a little bit," she said. "It is good for the kind of back injury you had and for the cold you have inside, the coldness of your internal organs. It will warm them up and it will help build muscle around the spine."

She explained that in cases like mine the body creates inflammation in the area of the strain or injury, but that inflammation can cause pain as it presses on the sciatic nerve bundle."

She paused and said, "I hope I am using the right vocabulary."

I told her I had heard it described the same way in Canada.

She wrote out "gou rou" and the Chinese characters with the word 'dog' beside it, so I could show it to someone in a Korean restaurant.

"How about lamb?" she said.

I told her I had eaten lamb but wasn't a big fan.

"Yang rou," she said. "Not quite as good as gou rou, but also good for you."

She wrote that out for me, too, with the Chinese characters to go with it.

She gave me a little Chinese lesson, too.

"It's not like English with many letters making one word," she said. "One character is one idea and you can put characters together to make words like dog meat or lamb meat."

Just before I left, I showed her the belt I got, based on her recommendation.

On an earlier visit, she strongly suggested wearing something that would keep my back, butt, and thighs warm. At a pharmacy in my neighbourhood, where they spoke only Chinese, I bought a back-warmer with velcro to hold it in place like a wide belt. Communication was difficult but it improved after I said the magic word.

"Bequin?" I said. "Do you know the name Bethune? From Ca-na-DA?"

I tried it several times, changing the pronunciation a little bit, each time.

"Ah, Bet-to-win," one of the women said.

So I gave her one of the Bethune buttons I had made up before I left Canada.

I gave one to the other woman working in the little store, too.

The two of them could not help me enough, after that.

They showed me a black belt that had eight magnets under a thin cover at the spot that would go over the small of your back and indicated that getting the magnets a little wet would make it get quite hot.

Dr. Wang approved of the belt. "Very good," she said. "The heat and the magnets will both be good. Also, just wearing the belt will give you some information you need, to take care of your back, while it is healing. It will remind you not to twist and turn and lift heavy objects."

"Hel how?" I said.

"Yes," she said. "Very good."

She walked me out of her office, down the hall, and to the stairwell, then leaned over the railing to say good-bye.

I told her I felt great and demonstrated my ability to take the stairs without using a railing or taking one step at a time or doing anything out of the ordinary.

"Yeah!" she said. "Bye-bye!"

"Thank you!" I said. "I feel great!"

"Great!" she said.

"You're the best!" I said.

"Okay," she said. "Bye-bye!"

5. Appointment #5

I was almost late leaving for acupuncture, so I walked fast to make the train, with no pain, and I was happy to discover that I could very quickly without experiencing any pain.

I caught the train and got a taxi guy I had before so he got me there fast and I arrived early, so I wrote for another half hour, while I was waiting for my appointment time. I wrote another section of a novel that I dreamed up while I was on the acupuncture table.

While I was getting needled, I was thinking about Dr. Norman Bethune and wondering about what his life was like when he was in China, and then I had a dream about traveling in time to meet Bethune while he was working with the Eighth Route Army during the war against Imperialist Japan. It was quite vivid and it left me with a strong desire to write it all down.

When Dr. Wang woke me up, to take out the acupuncture needles, and move on to the next part of the treatment, I felt groggy, as though I had been deeply asleep.

"That's good," she said.

I had a great time with the doctor, as always, talking about food, China, healing, and other things.

I showed Dr. Wang, pronounced Wong, that's right, my legs and stomach in my new long johns and then without my long johns as my legs were thinner but more muscled, and the calf muscles are carved like never before, and she put her hand on my stomach, saying that's very good, very flat, you look good, healthy and strong, with a lot of energy. She said she really likes me, a few times, because I am so full of energy, happiness, the love of life, and I love China, Dalian, Kaifaqu, the local food, acupuncture, healing, and everything.

I told her things had really changed over the past week so my feet no longer feel pins and needles plus numbness, they now feel full, like overstuffed sausages, like a water balloon about to burst, like I don't know what, like they are swollen, but they do not look swollen or anything.

She thought about it for a while and finally said, after asking me lots of questions about when and how, she said I've heard of this before but it doesn't happen except when someone has had nerve damage and usually the doctors say that's it, once there is nerve damage the nerves are dead, but sometimes they grow back, rarely, but when the blood supply to the area returns, it feels like the area is swollen, about to burst, but shows no signs of being swollen.

So, she concluded, I'm one of those rare cases and the healing work is working even though I haven't tried gar rou, yet!

Acupuncture went better than ever, but it could still be better, she says, as I will soon be able to feel the flow of qi like electricity from needle to needle, but it felt good, the time went fast, my nose didn't run, another good sign, and she didn't burn herbs in a box this time, just used the heat lamps.

Next she used her finger needling technique while walking around with me and usually I felt pain doing this, sometimes quite a bit of pain, but this time there was NO pain.

While we were waltzing around the room, I asked her, How long does it take for a nerve to regenerate?

She said, The length of time it may take for a nerve to regenerate depends on the exact nerve, the type of damage done, and if it is possible for it to regenerate. Many nerves do not regenerate, but it is the nerve endings that regenerate. Peripheral nerves can regenerate themselves if they are not completely destroyed. They can regenerate at a rate of 2 mm per day on the smaller peripheral nerves and 5 mm per day on the larger peripheral nerves. Regeneration of the nerves is not a speedy process, but is something that takes a long period of time.

I was 40 minutes early for my apartment so the doctor took me in early and then she kept me late as we were talking about Traditional Chinese Medicine and healing. The appointment lasted about three hours!

She made a date with me, to go to seven museums, close to where I live, on Feb. 1, the second day of New Year.

She said, You don't have to lose any more weight! You look good right now! Perfect proportions, strong, tough, lots of energy, good chi.

I told her I felt unwell the day before, after eating Western food. I went to a Subway and got a roast beef sandwich.

She said, It's too rich! Stick to Chinese food!

After the three hour appointment, I caught the bus to Kaifaqu, got off at the Subway stop, but did not go in, and walked down Jinma Lu to get my fave street food, again, which is bacon and egg wrapped in a thin pancake, with 2 sauces, plus lettuce, and it really hit the spot.

I walked all over Kaifaqu Station without ANY pain in my back, butt, legs, or feet. My feet had a lot of new sensations, which are hard to describe, but it is not pain, it's not tingling, not numbness, it's something else.

It felt a little like your feet feel after you've been skating for an hour with laces too tight and then the blood rushes back in .... But not in a bad way ....

It's hard to describe something you've never felt before.

We talked about acupuncture and fibromyalgia. She said it is REALLY good for fibro.

We talked about her qigung course, which was part of her medical training, for certification as a doctor, and about her instructor, who was gifted and could SEE chi ... and direct his own chi and direct his patients' chi ....

She said, It's only a little bit mysterious as Western science has been able to detect and document qi, the meridians, the points, just as they were documented 2,000 years ago in China.

We talked about the roots of TCM traditional chinese medicine in the big 3 philosophies: Confucianism, Taoism, and Buddhism, close to the heart of Chinese culture, as well, all these many centuries.

She said that is where the idea comes from that the human body is a microcosm of the universe.

"How is the human body a microcosm?" I asked her.

She said, "The body is a reflection of the universe as the pointed bones are the earthly manifestation of stars and mountains, the hollows are lakes and ditches, the meridians or energy channels are rivers, the organs are elements, and our internal weather (health, emotional stability, spiritual harmony) are influenced by the heat, cold, and damp of Earth as well as the wind and thunder of Heaven."

Dr. Wang said I am tough because fire cupping doesn't hurt.

She pricks me first and then puts the cups on the area in order to suck out the stale, dark, blood.

She says my injury that caused nerve damage stopped blood circulation as the muscles push old blood back to the heart and those muscles were damaged when the nerves were damaged.

Some of my muscles were shut down when the nerves shut down, but now my circulation is returning to normal. And that's why my feet feel funny.

Acupuncture has been a major part of primary healthcare in China for the last 5,000 years, she says. It is used extensively for a variety of medical purposes ranging from the prevention and treatment of disease, to relieving pain and anesthetizing patients for surgery. As in many oriental medicine practices, the emphasis of acupuncture is on prevention. In traditional Chinese medicine, the highest form of acupuncture was given to enable you to live a long, healthy life.

The earliest written account of acupuncture is found in the Nei Jing (The Yellow Emperor's Classic of Internal Medicine). This document is believed to be from around 200 BC and is one of the oldest comprehensive medical textbooks. Pien Chueh, a famous physician of the fourth century BC, used stone acupuncture needles, moxibustion, and herbs to bring a prince out of a coma. The Chinese still celebrate his birthday every year on April 28th.

The idea of harmony and balance are also the basis of yin and yang. The principle that each person is governed by the opposing, but complementary forces of yin and yang, is central to all Chinese thought. It is believed to affect everything in the universe, including ourselves.

Traditionally, yin is dark, passive, feminine, cold and negative; yang is light, active, male, warm and positive. Another simpler way of looking at yin and yang is that there are two sides to everything - happy and sad, tired and energetic, cold and hot. Yin and yang are the opposites that make the whole. They cannot exist without each other and nothing is ever completely one or the other. There are varying degrees of each within everything and everybody. The tai chi symbol illustrates how they flow into each other with a little yin always within yang and a little yang always within yin. In the world, sun and fire are yang, while earth and water are yin. Life is possible only because of the interplay between these forces. All of these forces are required for life to exist. See the table below to understand the relationship between yin and yang.

The yin and yang is like a candle. Yin represents the wax in the candle. The flame represents the yang. Yin (wax) nourishes and supports the yang (flame). Flame needs the wax for its existence. Yang consumes yin and, in the process, burns brightly. When the wax (yin) is gone, the flame is gone too. Ying is also gone at that time. So, one can see how yin and yang depend on each other for their existence. You cannot have one without the other.

The body, mind and emotions are all subject to the influences of yin and yang. When the two opposing forces are in balance we feel good, but if one force dominates the other, it brings about an imbalance that can result in ill health.

One of the main aims of the acupuncturist is to maintain a balance of yin and yang within the whole person to prevent illness occurring and to restore existing health. Acupuncture is a yang therapy because it moves from the exterior to the interior. Herbal and nutritional therapies, on the other hand, are yin therapies, as they move from the interior throughout the body. Many of the major organs of the body are classified as yin-yang pairs that exchange healthy and unhealthy influences.

Yin and yang are also part of the eight principles of traditional Chinese medicine. The other six are: cold and heat, internal and external, deficiency and excess. These principles allow the practitioner to use yin and yang more precisely in order to bring more detail into his diagnosis.

6. Appointment #6

Today she noticed that I'm losing weight so I said, I give all the credit to you. But she said, You get 20 per cent of the credit, your new diet and Chinese food gets 20 per cent of the credit, and acupuncture with herbs and moxibustion gets the credit.

She said, Acupuncture for your back uses the same channel as acupuncture for weight loss, so that is what is happening.

No heaving lifting, she said. And no twisting your back quickly, especially while holding anything, even a textbook, until the muscles around your spine get built up again.

We get our patients to wear a back belt and it gives them some support but it gives them a lot of information and reminders about how to take care of their backs.

Bend from the knees, not the back, and keep your back warm.

Your back belt, with the magnets, and heat, will be good for your back.

She put pins in some different places. She added the hips, on the outside, to the usual line-up of pins down the backs of my legs and the insides of my knees and calves.

She changed the position of the cups, as well, moving them from the middle of my back to the right side.

She measures carefully with her hands to find the best location for the cups.

Before she attaches the cups, she pricks me, then apoligizes. When the cups are in place, she taps on them. Sometimes she taps a few times and sometimes she taps a lot of times.

We just want a few drops of blood, she says.

Dark blood is not good. We want to get rid of that.

Cupping therapy is an ancient Chinese form of alternative medicine in which a local suction is created on the skin; practitioners believe this mobilizes blood flow in order to promote healing, the doctor told me. Archaeologists have found evidence in China of cupping dating back to 1000 B.C.

Fire cupping involves with the soaking of a cotton ball in alcohol. The cotton is then clamped by a pair of forceps and lit via match or lighter. The flaming cotton ball is then, in one fluid motion, placed into the cup, quickly removed, and placed on the skin. By adding fire to the inside of the cup, oxygen is removed and a small amount of suction is created.

Massage oil may be applied to create a better seal as well as allow the cups to glide over muscle groups in an act called "moving cupping".

For the first few sessions of fire cupping, the cups were just placed in one location. After a few sessions, the "moving cupping" technique was used.

Dark circles may appear where the cups were placed due to rupture of the capillaries just under the skin, but are not the same as a bruise caused by blunt-force trauma.

According to traditional Chinese medicine, cupping is a method of creating a vacuum on the patient's skin to dispel stagnation — stagnant blood and lymph, thereby improving qi flow — to treat respiratory diseases such as the common cold, pneumonia and bronchitis. Cupping also is used on back, neck, shoulder and other musculoskeletal conditions. Its advocates say it has other applications, as well.

The technique that Han (Jackie Chan) used in Karate Kid is called "Fire Cupping."

Fire cupping or simply cupping is a form of traditional medicine found in many cultures worldwide. It involves placing cups containing reduced air pressure on the skin.

The Karate Kid 5, as it's known, or The Kung Fu Dream, produced by Will Smith and starring Jaden Smith, also features fire cupping.

Jackie Chan is the kung fu expert and fire cupper in the 2010 Chinese-American martial arts action drama and remake of the 1984 film of the same name. It is the fifth installment of the Karate Kid series, serving a reboot.

After fire cupping, I felt as though I had been rebooted!

After acupuncture with moxi, followed by fire cupping, she put one needle just below my nose, and tapped it a few times.

That one stung a bit.

While that needle was in place, she said I could get my things on and then we would do some exercises.

She worked on my right hand, using both of her hands, digging her thumbs into the side of my hand.

She invited me to stand up, if it wasn't painful, and then to walk with her, forwards and backwards.

She never asked me to walk backwards before.

I see people walking backwards on the sidewalks and in open areas, I said. And she said, They are doing the same thing. It's good for your back muscles.

While we walk and she presses points on my hand, at the side and in the middle, on top, one hand and then the other, she asks me about pain and pays close attention to my responses and to my face.

The pain comes and goes. She asks me show her precisely where and then presses on that spot.

Just here? she says. Not here, or here.

It's the spot at the back of my left leg, at the top, and not above that or below that, she notes.

Very interesting, she says.

She digs her thumbs into the back of my hand and asks me to bend forwards and backwards, a little bit.

The pain comes and goes. It is not a big pain. It is a tightness, rather than a biting pain.

After twenty minutes, she invites me to sit down, and the pain or tightness quickly passes. I walk out feeling good: pain-free, loose, and energized.

I feel as though I fly down the stairs from the second floor to the first, down the hallway, out the door, and down a short set of stairs to the courtyard.

Walking to the bus stop is effortless.

Waiting for the bus, I take care to sit correctly, with a bit of a curve in my back, rather than slouching, and when I get a seat on the bus, I do the same thing.

From the bus to a taxi to the train, again, is completely pain-free.

In the afternoon, at work, I invigilate a three hour exam. To start and end the exam, there are a lot of stairs, as we use a computer room on the fifth floor.

After getting the students started, walking around to log them on to computers, there isn't much to do except watch for cheaters. I walk around the room, doing walking meditation, and I stand in different spots, doing a light tai chi or qigong exercise: shifting my weight from one foot to the other, feet slightly apart and at a 90 degree angle, turning the shoulders to my body faces one way and then the other, in alignment with my feet.

Three hours passes quickly.

I sit a bit and write poetry.

That means I am feeling good.

After work, I head home, sit at my desk while checking e-mail and social media for an hour or two, then make a snack, and head for bed early, listening to soft music.

Around midnight, I get up, after a few hours of deep, restful sleep, to get ready for bed. But first I go online for e-mail and so on for a couple of hours.

My feet feel fine and my back feels good, too. My legs are a little stiff but they loosen up as I dance to the music.

Dancing means I am happy, too.

After a couple of hours at my desk, sitting and typing, I feel some sensation in my lower back. On the pain scale, it barely shows up.

At 1:30 a.m., I go back to bed.

7. Happy New Year

Anne Millyard, Canadian publishing legend, asked me what I was learning about China while inside China, so I told her what happened to my during Spring Festival, which we in the West call Chinese New Year.

My doctor, who practices TCM, or traditional chinese medicine, and her husband who also practices TCM, took me out to celebrate Spring Festival / New Year and we had a blast.

I've had a lot of acupuncture, here in China, and my doctor is teaching me a lot about TCM as we go.

The three of us went to see six attractions in six hours.

We saw a planetarium, the exhibit about the body which came from Germany and is touring the world (it was at the Science Centre in Toronto), an archeological exhibit, a wax museum, a button museum, and a trompe l'oeil exhibit.

Each one was huge and occupied a 3 story building, all joined together. As we walked, we talked about the roots of TCM, which are in Taoism, Buddhism, and Confucianism,
just like Chinese culture.

The body is like the universe, they believe. And we saw the at the planetarium, inside the body at the body exhibit, and we saw Chinese culture through archeology and wax.

And that's what I think the Chinese people are all about. Right now, there is a consumer revolution happening in China and we could see that by looking at all the new cars in
the parking lots and on the new highway. And the 200,000 buttons in the Mao museum were a reminder of the communist era in China.

While we were taking a break for lunch, eating a jawza or two, a TV crew came over because they wanted to interview the westerner to find out what he thought about Dalian in wintertime, especially the area we were in, which is JinShitan.

The docs told them that I wasn't just some westerner, I was an author and educator from Canada, and I was from the birthplace of their hero, Norman Bethune.

They got excited about the interview after that!

They bought my lunch and then turned on the camera and let me go on and on. I raved about the Black Mountains and the Yellow Sea, the beautiful sunrise that turns the sea yellow, the attractions in the area, and how the planetarium plus body exhibit plus archeology and wax echoed the philosophy which is at the heart of Chinese culture: the idea that the body is like the universe.

So that's what I'm learning about China and the Chinese people inside China: They care about what Westerners think, they still love Bethune, and they have an ancient culture rooted in Taoism, Buddhism, and Confucianism, never mind communism, and they remember when China led the world, and they expect their big country to do it again.

They are not imperialist and their war against imperialism is far from forgotten. But they are willing to lead and to help the world, as they have in the past. They believe the world needs them to take on this role as the American empire goes the way of the British empire.

I was pain-free for three days in a row -- a new record.

My doctor and her husband, also a doctor of TCM (traditional Chinese medicine), took me to see 6 museums in 6 hours, and wouldn't let me pay for a thing -- Chinese hospitality, they said. And they said, Happy New Year! Welcome to China!

The museums were not museums, that's just what they called them. there was a planetarium, an archeology exhibit, that body exhibit that was at the toronto science centre, a wax museum, a gallery of trompe l'oeil (pics on FB), and the Mao buttons, each of them two or three floors of a big building

The two physicians watched me walk all day and observed my lack of pain on stairs, standing, walking, for six hours.

At noon, at a gourmet food station, a camera crew came along and said they wanted to interview the white man, so I gave them Bethune buttons, and my docs explained that i'm a famous writer from the west, so they went crazy and did a big interview, which would be on Dalian News, on DLTV, at 6:30.

My interview was aired on DLTV, the number one TV channel in Dalian, on Dalian News, the big show here, at 6:30, which is prime time here as it is at home.

I tried to be concise on DLTV and speak in soundbytes, but sometimes I give long, complex, answers to simple questions!

8. Healing Prediction

Today at acupuncture the doctor told me that if my healing continues, my back is pain free for 100 days, I'm cured, so we have to watch for the next 50 days, and then it's clear sailing. She said I should be confident about that. the numbness and tingling in my feet is superficial now and almost out of the 4 outside toes (that sounds like a lot of toes). she says the peripheral nerves are regenerating and that happens rarely, but sometimes

Thanks to the two doctors, I feel more confident with street food, and know what more things are, so now there are a lot of great options available for me everywhere I go.

First I learned where all the western things are in Dalian -- restaurants and stores for food -- but now I know how to speak "taxi Chinese" and feel comfortable walking into any store, restaurant, whatever, and communicating with my small but growing vocabulary.

"Taxi Chinese" means just enough Chinese vocabulary to communicate with taxi drivers.

While going for acupuncture every other day, I felt very creative and I was very productive. I got a lot of writing done.

My new novel was rockin' -- Bethune Returns To China. I had two new poetry books in the works. I became a senior reviewer for Tripadvisor with lots of articles published online, which I planned to bring together in a book called Dalian Close-Up, the sequel to Far Away, Dalian, Far Away. And I was working on a non-fiction book called Intro To TCM: acupuncture for aching backs, weight loss, increased energy, and over-all health.

I found a Taoist temple that's 2000 years old and built around a waterfall and asked the doctor about it. She asked around about them and phoned them, looked them up online and printed off a map for me, and found out a few things about the place.

She told me nobody answered the phone, they had an answering machine, so maybe they weren't open in the winter, but she thought they should be. Even so, she said, it might be better to visit the temple in the spring or summer, when the weather was better, as the waterfall might be frozen, and I would be in better shape for hiking.

There's a lot of stairs to climb and it could be hard on the back, especially in cold weather, so it might be better to wait until I was more healed, to make sure I didn't re-injure myself.

The day after we toured six tourist attractions in six hours, I returned to the clinic for acupuncture, as usual, but this time I got very different results.

I arrived early, so I sat in the waiting area, pulled out my pen and notebook, and did some writing practice, as I often do.

As soon as I started writing, the women in the waiting area came over to see what I was doing. I thought that perhaps they had never seen anyone writing in English before. I watch people writing in Chinese the same way.

Since they all worked at the hospital, I asked them if they knew Bethune.

"Bequin?" I said. "Ca-na-DA?"

They discussed it amongst themselves and then said, "Bequin!" nodding their heads.

So I gave each of them a Bethune button.

They smiled and said thank you, or "Xie-xie", and then left me alone to work on my writing.

Dr. Wang walked up the stairs to her office a few minutes after that.

We began by talking about our six hour long tour of the attractions in JinShiTan, the day before.

First, I gave her a New Year present: I gave her a box of vacuum packed tea. It was a little box of individual-sized servings. Along with the box, I gave her a card. It was a

Bethune card, from the Bethune House in Gravenhurst, which they suggested I buy and take with me when I went to China.

Judging by the doctor's reaction, they were right.

She did not open it right away. She waited until after she had needled me and she had a few minutes to do other things.

The card was printed on paper that looked hand-made and it had no writing on it, just a picture of Bethune, in gold, on the outside. It was a stamp made from a famous photograph of Bethune when he was in China.

He looked a little rough, I thought, but a lot of people liked that picture of Dr. Bethune.

Inside, I wrote her a little note, calling her the best doctor and saying thank you -- xie-xie -- for needling me and teaching me so much about TCM.

She said, My husband and I calculated the time. It was about six and a half hours, including lunch, so it was over six hours of walking. And lunch included standing up for the TV interview.

Are you okay, after that? she said. I felt bad, later, because we made you walk so long and climb a lot of stairs.

No problem, I said. I walked home after that. And I haven't felt any back pain, or any sensation in that area, for three days.

Three days, she said. That's very good.

Even after all that walking and all those stairs.

No problem, I said again.

Good, she said. That's very good.

Lying on my front, getting needled, the way we always begin, felt different as I was much more sensitive to the needles than ever before. In particular, the spots at the top of my right leg were very sensitive. When the doctor popped in the needles, I jumped and yelped.

Oh! Sorry! she said. You must be much more sensitive than before!

That's good, she said. It means you are healing.

My left leg was not so sensitive.

When all the pins were in, I thought I could feel that they were all connected, or as though there was a line of electricity connecting the needle points going down my legs.

The twenty minute session went by quickly, as it had the last few times.

Okay, the doc said. Roll over onto your back.

Instead of more acupuncture, she did a lot of finger needling. She stuck her fingertips into different spots on my left hand and then my right hand and asked me how my feet felt.

First of all, she asked me if my hands hurt, if her needling fingers felt painful, and I said, No.

Good, she said.

After a few minutes, I said, I have to tell you, my feet feel quite different. That numbness and tingling has gone from my big toes and my smallest toes.

How about the rest of your feet? she asked.

It feels superficial, as though it's only close to the surface of my skin, on the bottom of my feet, instead of going all the way through.

That's very good, she said.

After more finger needling, she made an announcement, or a pronouncement: "You are healing and should feel confident that you will heal," she began. "Peripheral nerve damage does not always mean the end for the nerves. Sometimes they regenerate. Not always. Not often. But sometimes."

"Great!" I said.

"In your case," she added, "it is happening."

"If you continue to be pain-free for another three days and another three days and so on, for about six weeks, or around fifty days, we say you are on your way to recovery.

"It takes one hundred days, we believe, she said. And after the first fifty days, it is a sure thing.

"So, we will watch for the next month and a half. But I believe you will be better by the time you have to fly to Canada for the summer."

I was so happy leaving acupuncture that morning that I decided to do something different on the way home. It was a holiday, the day after New Year, so the buses were on a slow schedule and there wasn't much traffic. It was Groundhog Day in Canada but not in China. I decided to grab a cab and go a different route.

Instead of running back to KaiFaQu, as I always do, I went to a different chingway station. I flagged a funny-looking taxi: a red, three-wheeled, Chinese car, which I barely fit in. I had never seen a taxi like that, before.

Instead of saying Kaifaqu, I just said, "Chingway", and he took me to the closest train station.

It was very interesting going a different route. It cost less, too. When I got to the train station I discovered a couple of things.

The good news was that there was a collection of taxi cabs there, on a slow day, so there would probably be taxis there on normal days, too. That meant my trip from JinShiTan to the integrative health clinic could be cut down. Instead of going allthe way to KaiFaQu and backtracking to get to the clinic, which took an hour, and cost 30 Rmb for a black taxi or 20 Rmb for a blue taxi, it would take about half an hour and cost about 10 Rmb.

The bad news was that there was a long flight of stairs and no elevator at that train station. I counted 52 stairs.

But I climbed those stairs without any problem.

No back pain. And my feet felt fine.

I felt ecstatic!

9. Dr. Wang Says

I asked Dr. Wang about Confucius.

She laughed.

I told her I had studied with a Zen Buddhist and a Taoist back home and both Buddhism and Taoism were popular in the West, but what about Confucianism.

She said, Confucius is one of the most quoted personalities ever. He is so popular that there is a special "Confucius says …" joke-selection, I mean who can say to have this kind of achievement .

Confucius, whose name literally means "Master Kong", lived 551-479 BCE. He was a Chinese thinker and philosopher, whose teachings have deeply influenced not only Asian thought and life. He presented himself as a "transmitter who invented nothing" and he really pointed out the importance of learning, which is one reason he is seen by Chinese people as "The Greatest Master".

One of the best known sources of Confucius are The Analects, a collection of his teachings, which was compiled many years after his death. A fountain of extremely mindful and wise Confucius quotes springs from these ancient descriptions.
Many of them are universal and timeless in their beautiful and simple truth and they are as valid today as on the day they left Confucius' mouth. Here we take a look at 10 of the most inspiring quotes by Confucius.

Confucius says …
1. "Never impose on others what you would not choose for yourself."
It's the "Golden Rule".
2. "Real knowledge is to know the extent of one's ignorance."
3. "I hear and I forget. I see and I remember. I do and I understand."
4. "Everything has beauty, but not everyone sees it."
5. "The Superior Man is aware of Righteousness, the inferior man is aware of advantage."

6. "Wheresoever you go, go with all your heart."

7. "Our greatest glory is not in never falling, but in getting up every time we do."

8. "He who learns but does not think, is lost. He who thinks but does not learn is in great danger."

9. "He that would perfect his work must first sharpen his tools."

10. "If you look into your own heart, and you find nothing wrong there, what is there to worry about? What is there to fear?"

10. Downtown Dalian

I had a good trip downtown after acupuncture in Kaifaqu. I felt inspired to go to Dalian's huge downtown area to find the big bookstore.

In china there is just one chain and one location per city, apparently. My colleagues told me the only place to get books in English is in Beijing, but I discovered our book store has a section for us. It used to be the state's propaganda department, during the cultural revolution, but now they sell books and related stuff (magazines, translators) from around the world.

It's huge! Nine floors in a big building full of books!

The books I bought cost about the same as books in Canada, 20 bucks each, on average.

Dr. Wang spent an hour online getting me maps, directions, phoning to make sure they were open, and writing useful phrases in Chinese, such as "books in English". None of it was necessary but I certainly appreciated her efforts.

She is crazy about street food in China, especially in Dalian, so I'm feeling more adventurous in that department.

When you get off the LRT train downtown, you walk under the Dalian high speed train station, that connects us to Beijing, and the walkway under the big station is a lot like Beijing, they say, as traffic is blocked off so there's a six lane street full of people as the street is lined with vendors selling everything, especially food.

I bought stuff on the way into the city and on the way out again!

I found a Subway, selling North American food, and got excited for a second, but then went for the Chinese street food instead! It's better, tastier, cheaper, and not too rich, like food in the West.

Acupuncture is simpler now, no moxibustion, just needles, lamps, and finger needling, no walking while needling, so I asked why. The doctor said it is because I'm almost healed! The day before was good but I had some sensitivity in my lower back after 3 days with no pain, so I was a little worried.

I walked to the market in Manjiatan in record time, 15 minutes instead of 30, but I had a touch of pain in one leg.

So I was happy but not ecstatic going to see the doc on Thursday.

I made it door to door in 45 minutes, another new record.

We had a great time, as always.

I gave her a book from the book store she helped me find. She looked it up online, phoned them to make sure they were open during the Spring Festival holidays, and printed me a map, then worried because she got lost when she looked for the bookstore.

The book I got her was by Yan Mo, China's Nobel Prize winner. She was very touched. My first novel in English, she said. Foreign books are very expensive! You've given me so many gifts!

I've given her a button, a card, some tea, and a book. That's it!

I told her she had given me far greater gifts, fixing my back and explaining so much about TCM.

She read the cover blurbs. He is compared to Milan Kundera, she said.

Yes, I said. I love Yan Mo and Kundera. He's a Czech writer in exile, living in France, the author of The Unberable Lightness of Being. A great novel turned into a very good movie.

She said, You know, Yan Mo was famous in China before he won the Nobel Prize. He won the top Chinese prize for literature.

Yes, I said. He was a best-seller in China and made a lot of money. He won The Mao Dun Literature Prize, awarded by the Chinese Writers Association, every four years. He won it with his novel called Frog. Or Frogs. It's set in the time Bethune was here in China and it describes the Eighth Route Army, which he was attached to as a doctor.

Oh! she said. You know about Chinese literature! That's very good.

I don't think Frog has been translated into English, yet, I said. My favourite book by Yan Mo is Life And Death Are Wearing Me Out.

Your hands are so hot! I said.

A lot of energy, she said.

She needled me and we talked about needles and needling.

I asked her if I could ask her a lot of questions and she said, Yes, I like being asked questions, and I like your questions. It shows you are interested and you are very intelligent.

When it comes to acupuncture, is it the location of the needles, or the needling; that is, the way you move the needles.

She said, You have to put the needles in the right places, you have to know all the places, and you have to know which of those places will be the best, at the time. But moving the needles is the most important part. You have to feel the energy and use your hands or fingers to twist and turn the needles, to get the right angle and depth.

She said, I had to do a very big exam for that. There was a practical exam and a written exam. They watched your technique and the way your patient responded. It was very hard to write about because a lot of Chinese characters were required to explain how to use those little needles.

I asked her about working in the U.S.A.

She said she was in California for four years, working with one of her old professors, from China. She did rounds, observed by interns; she did rounds with interns participating and listening to her advice; and she gave lectures with demonstrations to classes of 30 or 40. She would demonstrate on a patient and then talk for a long time, explaining what she did.

She said the most important part, for her, was letting the Americans know that what she was doing was not mysterious or magical, it wasn't just ancient techniques practiced out of tradition, it was a medical treatment that worked and could be verified by science. She said Western diagnostic tools show chi and the meridians and the acupuncture points.

The latest electronic tools with lasers and computers identify the same things as Chinese medicine over two thousands years ago.

I asked her about finger needling, or tuina, and she said it is part of TCM with acupuncture, moxibustion, herbs, and nutrition.

What about tai chi and qigung? I asked.

Yes, she said.

Using these things is not like adding one plus one plus one, she said. It is more like multiplying two by two by two. The effect is expanded exponentially.

How come you used acupuncture with moxibustion and heat lamps when I first came here and now it's acupuncture without moxibustion and now with fire cupping plus tui-na, or finger needling, and the heat lamps, plus information about nutrition. You are healing, she said. Moxibustion was needed in the early stage and tui-na is effective now now that you are much better.

Heat is very important, she said.

She said that in California, people hurt their backs playing golf and going hiking in the mountains. In China, people hurt their backs sitting in long meetings. I said, People in Canada hurt their backs shovelling snow.

The worst thing you can do for your back is shovel snow in the cold, lifting and turning heavy loads. You should keep your feet, legs, and back very warm. It's good for the spine and discs and muscles of the back.

I told her I hadn't been able to find the thick socks we wore in Canada in the winter. She said she would find someone to make them for me.

I also told her that my friend, who introduced me to her, flew back home to Canada during our break for Spring Festival and she was sick in bed the whole time. She flew to Vancouver, where the weather is like Dalian's, and then she flew to Ottawa, which is cold. She shook her head. Western medicine is very good for diagnostics and analysis but Eastern medicine is very good for healing without drugs or surgery. Western anti-biotics are very good for fighting bacteria but Eastern medicine, especially herbal tea, cooked by a hospital, are the best for viruses.

Yin and yang, I said.

Yes, she laughed.

Confucious say .... Integrative medicine is the best, I said. Put together the best of the East and the West.

Yes, she said. Integrative medicine is growing fast in the U.S.

I told her I bought a soup pot and made some great soup with tofu, soy sauce, chili paste, and bean sprouts.

That's very good, she said. You are eating like a Chinese person. Stay away from rich foods.

No frying, I said.

No deep frying, she said. French fries and other things made in the deep fryer are very hard on you. You can fry things in a little oil, no problem.

What about soup? I said.

Very good, she said. Hot food, hot liquids, very good.

Protein in the morning, I said, repeating her advice from earlier sessions.

A little protein in the morning. Make breakfast the main meal with a lighter lunch and a small dinner.

After needling me, she used fire cups on my legs, pricking me first, and apologizing for causing pain, then pulling the cups down my legs, along the meridians.

After that, she got me to flip over, slowly, and she did some tuina work on my feet and my hands, asking me to move my back around while she worked. Just move it a little bit, she said, and tell me if there is any pain.

No pain, I said.

Very good! she said.

She looked for tender spots on my feet and hands but could not find one.

She found one on my forearm and she worked on that.

I asked her about yoga and demonstrated the small yoga moves a physiotherapist had given me in Canada.

That is actually bad for your back, she said. But there is one move you could do to strengthen your back.

She demonstrated a move called "flying".

You lie on your front on a hard surface like a massage table and lift your four limbs. It looks like you are freefalling, I said. I like freefall. -- That's a writing technique I like to use and teach.

You should teach me to freefall, she said.

I should get you going on that book you want to write, I said.

I did not tell her I was working on this book. I wanted to surprise her with it.

I asked her if she cooked. She said yes, but she had an aunt who was a very good cook and who did a lot of cooking for her family.

I told her I bought fresh tofu and cooked it up with garlic and onions, plus some red paste and soy sauce, and it tasted great.

Very good, she said. I love that.

When you get tofu on the street, they use garlic powder and sesame seeds. I haven't been able to find garlic powder, yet.

I'll write the Chinese characters for you, she said. When you show it to someone in the grocery store, they will be happy to find it for you. Do you get groceries in Walmart, too? The store is so big and they have so many things, it's hard to find what you are looking for. But they are very good at finding things for you and they like to do that. That's something else I love about China, I said. In Canada, it's hard to find anyone who can help you in the big box stores like that.

What else are you looking for? she said.

What are the sauces they use for jian bin? I said. A lot of street food vendors paint on two sauces, with a paint brush. One is dark brown and one is dark red.

They use that in the food court outside Walmart? she asked me. Which one.

I described the place where I liked to get jian bin and she said, Yes! I know the one! I like that place, too!

I told her they were closed the other day so I tried another one and discovered great jaowza and an eggplant dish as well as a sausage.

You like all that? she said. You are more like the Chinese every day. -- Very good!

Leaving the doctor's office and the integrative medical clinic, I felt good. I took the bus to Kaifaqu and felt quite well. But as I walked from the bus stop to AnSheng Mall, I started to go downhill, a bit. My mood deteriorated. I felt as though I was coming down with a cold or something.

I wandered down the street between AnSheng Mall and Mykal, toward the Dalian New Mart Mall, and got a wiener on a stick. I took a dozen pictures of different things on the sidewalks. Some of the street food vendors had been replaced by a ring toss game. I walked along the street, looking at all the things being sold, and found some thick socks, so I bought a few pairs.

And then I went to my favourite street food place in the underground food court outside Walmart.

They recognized me, so they didn't let anybody butt in ahead of me, and took my order right away.

The short order cook held up two fingers, the way I do, indicating he knew my order.

I took the jian bin upstairs to the outside court and sat at a table to eat. I got a drink inside the mall, to go with the bin.

While I was eating, I watched people go by. Whenever I sit there, people stare at me, so I say, Niehau, and they say, Hello!

I get the impression they are people who rarely see foreigners and are hesitant to say anything but like to try a few of the words they have learned in English.

I saw a deaf couple, using sign language, and speaking without making any sound, and I wondered if it was different for Chinese speakers than English speakers.

I said "Hi" in North American sign language and they both smiled and waved in response.

Those things made me feel a bit better. I did some writing practice, sitting at the Coca-Cola tables, and got some more stares.

After that, I went to a few bank machines, looking for one that could handle an off-shore bank card, and found one up the street.

In the grocery store, I found everything I was looking for: garlic powder, a shaker of five spices, tangerines, and a few other food items.

The Mandarin oranges were not in the big display where they are usually located. They had oranges of different sizes there, instead.

I also bought a foot warmer, the likes of which I had never seen. It had a fuzzy, white interior, a blue blanket exterior, and a plug.

The doctor encouraged me to keep my feet warm, so I thought I'd try it.

I went home and tried out the foot-warmer while checking e-mail, Skyping with Cynthia, and playing Scrabble online. I didn't write a word, other than e-mails, but I had a lot of e-mails to return.

The new foot-warmer felt great!

11. The Doctor Who Bought Me Socks

In the waiting area at the integrative clinic, waiting for the doctor for a few minutes, I start writing practice, as always, when I have a little time to myself, but this time I was quickly surrounded by women who work in the big clinic. They watched the big white guy from North America writing in English, not Chinese, how strange. I like to watch people write in Chinese, in just the same way.

So, of course, I said the name "Bethune" to them. "Bequin?"

They smiled and nodded, so I gave them Bethune buttons.

They went away, happy, and let me write in peace.

I told the doctor about that and she said, again, that Bethune holds a special place in the hearts of many, many, people in China.

I also told her that I had discovered the meaning of the name "Bequin" and I had thought it was just something that sounded like "Bethune".

When I got my I.D. badge at work, it had a Chinese name on it, for me, but when I had it translated, I found out it sounded just like my name in English.

The doctor said a lot of people in China know the name Martin Luther King, so the Chinese characters for the name Martin are well-known.

She said Bethune's name was probably like that, too, but I told her I read somewhere that "Bequin" means "White one seeking grace" and that in the local dialect of the people in the area where he worked the same Chinese characters meant "one who was sent".

"Like a god?" she said. "I never knew that."

I told the doctor that a friend from Canada called me on Skype and said she was so amazed about the was I was shrinking that she wants to try acupuncture, but I told her that you have to go frequently, not just once, so she didn't know how she would pay for it, as it isn't covered by her health care plan, public or private.

The doc said, She can go once and then take herbs.

It's best to go every day or every other day and get acupuncture with the other types of Traditional Chinese Medicine: moxibustion, fire cupping, finger needling, herbs, nutrition, and energy exercises (tai chi and qigun), but she could start with acupuncture, to get things going, and follow up with herbs. That's a good way to lose weight.

I told the doctor that a filling popped out while I was chewing on the herbal formula she gave me, so she took me downstairs to see a dentist, and she acted as translator for us.

The doctor said it's important for me to keep my feet, legs, rear end, and back warm, as it would be good for my spine, muscles, nerves, circulation, et cetera. I told her about the thick socks we wear in Canada and said I hadn't seen anything like that in China, yet, but I was still looking.

She bought me socks and gave them to me at our next session.

"How much?" I asked.

"It's a present," she said. "You've given me so many presents."

She said they were the thickest socks sold in China.

She told me I was on 1 and DLTV on Friday night and she called the Dalian News to get them to send her a dvd of the show, so she could give it to me.

I told her that on Saturday, the day after the show aired, I was recognized on the street several times. People who did not speak English said, "Hi! TV! Hello!" So, I gave them Bethune buttons.

She said, "You are famous and you are kind. That's the way!"

The doctor said a lot of money was spent in the U.S.A. and China for studies on the efficacy of acupuncture with different groups getting needled daily, every other day, every three days, once a week, and so on. They determined that daily acupuncture or every other day was the best. And she said that TCM doctors everywhere said, You didn't need to spend any money on that; we've known those things for over two thousand years!

I asked her what she did to keep up her energy. Her hands are hot. She said she does energy exercises online for her own healing energy and that's what makes her hands so hot. And she promised to send me an e-mail with a link to the site.

At this session, she gave me acupuncture for twenty minutes, with heat lamps, and then had me turn over, for finger needling. She used her fingers to massage different spots on one hand and then the other and then one foot and then the other.

    We both noticed that I was more sensitive to the acupuncture needles and less sensitive to finger needling. I felt the qi circulating more and had far fewer tender spots on my hands and feet.

    "That's good," she said. "Those things mean you are healing."

When the acupuncture needles are in and she has manipulated them, they feel connected, as though there is an electrified wire going from one needle to the next, with a gentle, warming, current. It feels good.

    "That's good," she said.

    She asks if I can feel the energy going down my leg and out my toes.

    I imagine and visualize that happening.

After acupuncture, she uses the fire cups. Instead of placing them on my back, she massages the meridians at the back of my legs, tracing the meridian down from my thigh to my calf.

    "Too much?" she asks.

    "It's fine," I say. It feels good.

I asked the doctor about the "big three" movements that are at the heart of TCM and also the roots of Chinese culture: Buddhism, Confucianism, and Taoism. "Is there one you follow? Do you belong to a temple? Or do you maintain a connection to all three? Or did you just study them at university?"

She said she liked all three, had not been able to pick one to follow, yet, but her in-laws were interested in Buddhism, so she thought she might go with them, but had not joined any temple or group.

"Buddhists are happy!" she said. "They are open-minded and intelligent. I like the Buddhists sayings that sound so wise."

I told her I loved the time I spent with a Zen Buddhist monk but I was drawn more to Zen than to Buddhism and now I felt more drawn to Taoism.

"You have Taoist temples, here; in Canada, we don't have anything like that, that I ever saw."

She said she phoned the Taoist temple on Black Mountain, again, for me, but just got their answering machine, one more time.

"It's more beautiful when the grass is green," she said. "Maybe it would be best to wait until then."

We talked about the Olympics, a bit, as the 2014 Games were starting in Sochi, Russia, and I was watching online. She said Chinese Olympic athletes are national heroes in China. "If you win a gold medal, you might be set for life," she said.

I told her Canadian athletes had jobs, did fundraising, might get a parade, plus publicity. A few made some money through advertising endorsements.

I told her Canada's best skater was a Chinese-Canadian guy.

She liked that!

I told her China had some great skaters, too.

"Oh?" she said.

At an earlier session, we had talked about nutrition, which led to some talk about cooking. I asked her what sauces were used at the street food places. There's a dark red sauce and a dark brown sauce, I told her. They put it on tofu, on jian bin, and other things, including squid on a stick, I said. She said she would find out and get the names for me and she did just that. She went to a street vendor and asked about the sauces. She got the names and wrote them down for me, in English, Pinyin, and Mandarin, so I could take a note to a grocery store and show it to someone so they could find them for me.

She said that after she did that, she asked her aunt, who did a lot of cooking, and found out they use those two sauces at home all the time.

They come in squeeze-bags, she said, and they are very popular. One is a little sweet and one is a little savoury.

12. Still Healing

At this session, my legs were even more sensitive to acupuncture than the last time and my hands and feet were even less sensitive, or felt less tender when the doctor used her finger needling techniques.

I asked her if she was going easy on me, out of pity, as she had tortured me so much in the past.

"No!" she insisted. "I like torturing you!"

We were just joking around, of course; it's the opposite of torture.

She pressed a spot on the bottom of my foot, in the middle, and I was moved to say, "Hey, that feels good!"

"That's the kidney spot," she said.

She gave me a lengthy explanation about why it felt good. She said they compared the human body to nature, really see the body the way they see nature, and think of qi as a river running through the body. There is a spring at this location, she said, like a little waterfall.

I asked her about moxibustion as I had read online that sometimes the herbs are burned in a little box and sometimes garlic, ginger, or salt is used. She said that in very serious cases, the herbs were set on a slice of ginger, or on top of salt, depending on the ailment, but that in my case, that wasn't necessary, so she used the box, which is less direct.

I also asked her about the acupuncture hammer I had read about. She showed me one and explained it was used for superficial stimulation of qi as the needles were very short and did not penetrate the skin.

She asked me about the sensations I was feeling and I reported that there was less tingling in my feet and no pain in my back, legs, feet, or anywhere.

It had been twelve days in a row with no pain and that was quite a relief, I told her, as it was the longest pain-free period I had experienced in over a year.

The pain was never great, but it was a worrisome annoyance, and it felt fantastic to get rid of it.

"You are healing!" she said. "It's good!"

13. More Acupuncture

After acupuncture for two days in a row, I skipped a day, and was worried because I felt a bit of back pain and a little strain in one leg, on my day off, and the doc said we were looking for 50 pain-free days in a row as a strong indication of healing.

It was a cold day and I spent a lot of time writing, then went for a walk, despite the cold, as I wanted to go to the Great Market in Manjiatan to get some groceries and other things. I felt a little pain on the way up there, but it went away as I walked around the outdoor market and then the indoor markets.

It takes 100 days, she says, and after the first 50 days, it's clear sailing.

When I told her I had a little pain the day before, she said, That should not be a problem. Don't worry. If it just came and went and wasn't too bad, it should be okay. Yesterday was very cold, she said, and I bet you were writing for a long time and then walking outside in the cold.

That made me feel better.

Once again, I was more sensitive to acupuncture and less sensitive to finger needling.

Fire cupping was different, this time, as she cupped a spot on my shoulder, which she said was for the sciatic nerve bundle, as well as running the cups down my legs.

At the Manjiatan market, I ran into one of our teachers, who has lived in China for a decade or so, and he introduced me to his "chicken lady" -- the person who sells him chicken in the meat market.

"Chinese people prefer the meat that's close to the bone," he said, "so chicken breasts are cheap, as nobody wants them, except foreigners like us."

He bought a big bag of chicken breasts, chopped them in two, put different marinades on them, sealed them in freezer bags, and stored them in his freezer until he felt like a quick dinner with chicken.

That sounded good to me but when I looked at all the meat in that part of the market, I had a sudden desire to eat tofu for dinner. The North American style of eating, with a big meal at the end of the day, no longer appealed to me. I liked my new routine, with a big breakfast, smaller lunch, and tiny dinner.

I told the doctor that story and she said, "Very good."

She confirmed that people in China prefer the meat that is close to the bone, rather than chicken breasts, or big pieces of meat like that. And she recommended chicken feet.

That raised my eyebrows!

"You North Americans don't like that, but it is very tasty, she said. You should try it!"

I told her I would put it on my long list of different things to try in China. Somewhere near the bottom of the list, with blood on a stick.

"I haven't tried gou rou, yet," I told her. "I haven't seen it anywhere."

She said, "It's not easy to find. Only some Korean restaurants have it."

She promised to make some phone calls and find me a restaurant that prepared gou rou well and served it tastefully.

I told her I had shared her recommendation with some Canadians and they all said the same thing: "There is no way I am eating dog meat!"

And I told them I used her line on them: "Do you eat deer?" I asked them. "Is eating dog meat worse than eating Bambi?" I added.

Just for the record, I've been a vegetarian for most of my life. I tried deer meat once, as venison sausage, and I tried bear steak, once, as well. I might try gou rou, if it will help the healing process.

Once again, I thanked the doctor for all of her work on my behalf. And she said, "It takes two. The doctor and the patient have to work together as a team for healing to happen. You have to take the herbs regularly and look after nutrition while I do acupuncture and so on."

I asked her about tumeric, as it was something Dr. Weill recommended on a recent posting on social media. I told her who Dr. Weill was and asked her what she thought of him.

"He promotes integrated health care," I said, "and he's quite famous for it, in the West."

"Oh?" she said. "I will look at his website and see if what he says goes along with Traditional Chinese Medicine."

Tumeric, she said, was a herb in China, used for healing purposes.

I told her it was a spice in Vietnam, used in cooking, and Dr. Weill recommended it for cooking, as well, as it helped reduce inflammation.

"It works well for inflammation," she said, "but it has a small effect compared to the herbs you are taking. So, don't use it right now. Do two courses of the herbal formula, let them have their full effect, and then you can use tumeric after that."

I asked her about the holiday in April, on the fifth day of the fifth month, called Tomb Sweeping, as I read somewhere that it had something to do with Traditional Chinese Medicine. She said, "That is the day that willow branches are picked for Chinese Medicine. Some are hung on doorways, to purify the air and the environment."

My other question of the day was about Dr. Norman Bethune. We had talked about his name once before. I told her I did some more online research and discovered that his Chinese name, which sounds like Bethune, but has a "q" in the middle, Bequin, means "white one seeking grace".

She wrote out the three characters for be, qui, and in, and she said, "Yes, bei means white, as in a white person, and qui means seeking, and in means grace, so be qui in or Bequin could mean white one seeking grace.

"White is a word we use in China for Caucasians as we use the word Yellow for Asians, and so on. I know that in the U.S.A., that isn't done, it's considered impolite, or worse, but here we use White, Yellow, Black, and Brown without meaning any insult or anything like that. They are just colours, with no connotations. You are White and I am Yellow. Bethune was What. They found a name that describes him and sounds like his English name."

I asked about the regional dialect that translated the same name and characters, Bequin, as "one who was sent".

"Maybe they did see him as a person who was sent, like a god," she said, "as he saved the lives of thousands of people."

I told her that when he took his mobile hospital to some of the remote places in the mountains, they took a generator for electricity, and some of the people living there had never seen electricity or a white person before, so it must have been very impressive and godlike to have a white man appear with electric lights and watch him operate on soldiers and villagers in ways they had never seen or heard of before.

She didn't comment on that. She just nodded her head.

14. Energy Exercises

The doctor sent me an e-mail, as she promised, with a link to energy exercises we had discussed.

Hello Martin,
Here is the link to the Ba Duan Jin "eight sections"
http://v.ku6.com/show/9Q7G8Ttu39gpBFYBxOKOpg...html
Just have a look. This is very common for people to practice for improving energy and preserve health. Share with you!
Sincerely,

Dr.Wang

As a certified Qigung instructor who spent several years at a Zen retreat called The Zen Forest, in the Far East of Ontario, Canada, where we followed meditation with massage and then energy exercises as a regular part of the daily routine, I knew the Eight Sections Qigong as The Eight Section Brocade Qigung.

One tradition is that Bodhidharma (448-527 CE), a famous Grand Master of Chan (Zen), introduced a set of 18 exercises to the Buddhist monks at the Shaolin Temple known as the "Eighteen Hands of the Lohan" and his Shaolin Lohan Qigong (i.e., the art of the breath of the enlightened ones) "is an internal set of exercises for cultivating the "three treasures" of qi (vital energy), jing (essence), and shen (spirit)"

Some say the eight exercises were developed at the Shaolin Temple a few hundred years after the Bodhidharma died.

There are numerous versions, seated and standing, of Bodhiidharma's exercise sets.

I found a great link to a full description of the eight movements, with some history and theory: http://www.egreenway.com/taichichuan/esb.htm#Illustration.

DVDs of the qigong set are available online from many outlets, including Amazon.com and Walmart.

There are numerous versions of the set on YouTube. A good one is called Qi Gong: Shaolin Ba Duan Jin (8 Brocade Exercise).

15. The Five Elements in Chinese Medicine

After acupuncture with heat lamps followed by finger needling and fire cupping, the next time, the doctor told me about the five elements, also called "Wu Xing". She said, "The body is like nature and the the cycles of nature correspond to the human body.

The Chinese term "xing" means the process of one thing acting upon another. She said when I use acupuncture with heat and finger needling plus fire cupping, it is not like adding one plus one plus one plus one, it is more like multiplying 2 X 2 X 2 X 2.

Here's how it works:

Wood feeds fire. Fire creates ashes which form earth. Ii=nside the earth, metal which is heated liquifies and produces water vapour. Water vapour condenses and turns into water that nourishes the trees, or wood.

The five elements, in the body, work like this:

Fire is hot, ascending, light and energy, in the heart (yin) and small intestine (yang).

Earth is productive and fertile, for growth, and relates to the stomach (yang) and the spleen (yin). The stomach begins the process of digestive breakdown, while the spleen transforms and transports the energy from food and drink throughout the body.

Metal is a conductor. This element includes the lungs (yin), which move vital energy throughout the body, and the large intestine (yang), which is responsible for receiving and discharging waste.

Water is wet and it is descending, or flowing down. The water element represents the urinary bladder (yang), and the kidney (yin). Water sends fluids throughout the body, moistening it. and then it accumulates in the kidneys. The kidneys also store the essence, and serve as the root of yin and yang for the entire body.

Wood is strong and rooted. The wood element represents the liver (yin), and the gall bladder (yang). The liver stores blood and regulates the smooth flow of qi.

Interesting, I said. I'll have to learn more about that.

16. Jing, Chi, Shen

The next time, before acupuncture, the doctor told me about chi, jing, and shen.

I asked her about jing. I told her I had been reading about Taoism and TCM and came across the term jing. In the West, I said, a lot of people are now familiar with the term "chi" or "qi", but I don't think people are familiar with the word "jing".

I heard the term in a movie called Bliss and in books by Mantak Chia, a Taoist, working in Upper New York State, at some mountain.

The doctor grew serious and talked for a long time about jing, as well as chi, and she added in shen.

Chi, Jing, and Shen are known as the Three Treasures, in Traditional Chinese Medicine or the Three Jewels. They are the cornerstones in TCM as well as Qigong, and T'ai chi.

They are also known as Jing Qi Shen (pinyin: jīng-qì-shén). In English, we say ching ch'i shen. It means "essence, qi, and spirit".

Jing, qi, and shen are three of the main concepts shared by Taoism and Chinese culture.

In long-established Chinese traditions, the "Three Treasures" are the essential energies sustaining human life:

Jing is essence, Qi means vitality, energy, the life force; and Shen means spirit or soul.

Jing-qi-shen is used more than qi-jing-shen or shen-qi-jing.

If someone asks you how you are, how you are feeling, or how you are doing, you might say, "Jing-qi-shen" and give them the "thumbs-up" sign, if you are doing well.

She said, Chi is like air, as it's invisible, but it circulates throughout the body, like electricity. It is energy. It is related to Jing.

And Jing is much more condensed. It is stored in the kidneys. It is something you inherit from your parents. When you run out of Jing, you run out of life; in other words, you are dead.

Chi nourishes Jing and Jing creates chi.

Shen is your spirit.

All three are very important for your life and for your health.

When I taught Traditional Chinese Medicine in the U.S.A., and when I studied it at Beijing University, there was an emphasis on appreciating these three concepts and how they relate to the work we do.

I liked learning and teaching the philosophical side of Traditional Chinese Medicine.

In the West, many people know the term "chi" but not "jing" and "shen". In China, most people know "chi", "jing", and "shen", but they might not know a lot about them.

There was a time when everyone in China knew all about these concepts, but that was lost when Western or scientific medicine came to China, for a while. And now these things are getting to be well-known again. There are television programs that teach people about these things in an entertaining way so everyone understands what they mean and how important they are for maintaining your health and your life.

Jing, or essence, is the substance responsible for reproduction and regeneration. It is believed to be derived from two sources: the energy inherited from one's parents and the energy a person acquires in his or her daily life, from air, food, and water. Jing regulates the body's growth and development, and works with qi to help protect the body from harmful external factors.

According to tradition, Jīng is stored in the kidneys and is the most dense physical matter within the body (the opposite of shén, which is the most volatile). It is the material basis for the physical body and is yīn in nature, which means it nourishes, fuels, and cools the body. As such it is an important concept in the internal martial arts. Jīng is also believed by some to be the carrier of our heritage (similar to DNA).

One is said to be born with a fixed amount of jīng and also can acquire jīng from food and various forms of stimulation including exercise, study, meditation, and so on.

Jing and qi have a close relationship. In traditional Chinese medicine, they are believed to form the foundation for the shen, or spirit.

Shen refers to spirit, god, deity, the spiritual and the supernatural, as well as awareness, consciousness, et cetera. The primary meaning of shen is translatable as English "spirit, spirits, Spirit, spiritual beings; celestial spirits; ancestral spirits" or "god, gods, God; deity, deities, supernatural beings", etc. Shen is sometimes loosely translated as "soul".

Chinese people like to look after their health, she said. As a doctor, I know it is a lot better and easier to do preventative health care than try to fix a problem later on. It's much better to prevent a disease than get the disease and try to get rid of it.

Jing energy is the primordial energy unique to an individual that is passed to them at conception. Shen is the driving energy behind activities that take place in the mental, spiritual or creative planes. Qi is the most dynamic and immediate energy of the body.

We need all three for health and for life.

She said something that startled me, at that point. She said, "When you first came here, I asked you to show me your tongue, and I remember what it looked like, as it was pale in the centre, purple on the sides, did not have much energy, and had a thick coating.

My first impression was that it would take a lot of work and a lot of time to get you back to a healthy condition again. That was in December. But now it is mid-February and you have come a long way. We could now say you are "Jing-qi-shen"!

And she gave me the "thumbs up" sign, smiling broadly.

You have changed your diet, taken one course of herbs, and you'll take a second course, you've had a lot of acupuncture, some moxibustion and fire cupping, as well as finger needling, you get a lot of exercise, and you have replaced cold food and drinks with hot food and drinks. It has all had a good effect on you in a short time.

Your back is healing and will be healed within one hundred days.

Nerves heal more slowly, but your peripheral nerves, in your feet, are healing, too.

The (late 16th century) Journey to the West novel provides a more recent example when an enlightened Taoist patriarch instructs Sun Wukong "Monkey" with a poem that begins:

Know well this secret formula wondrous and true:

Spare and nurse the vital forces, this and nothing else.

All power resides in the jing, the qi, and the shen;

Guard these with care, securely, lest there be a leak.

Lest there be a leak! Keep within the body!

I asked the doctor more questions about "jing-qi-shen".

Jing, Qi, and Shen, are all Mandarin. In Cantonese, it is "Jing", "Hey", "Sun".

"What would it be like to be the opposite of "jing-qi-shen"?" I asked her.

She said, "Jing Qi Shen is essential for a human being. We cannot live without one of them. Traditional Chinese Medicine is all about preserving and improving the three to improve life."

She also said, With Jing and Qi, your Shen gets stronger. But at the same time, Shen is a reflection of the three and all three will brighten up and have higher density when your Shen is bright and strong.

With Jing and Qi, Shen is generated, or vitalized.

Shen is the spirit, which is an abstract word for a person's mental state and health condition. When one is tired and sick, the Shen is described as being dim and weak. When one is alive, awake, and active, we describe their shen as bright and vivid. It is not physical, but it is visible by the symptoms shown on the physical body. The best way to see if one has "Shen" is to look at the eyes. Sleepy eyes often indicate you are lacking energy; therefore, Shen is dim and weak.

She said someone in that condition, or that kind of shape, would find finger needling very painful at every point.

She reminded me that the first time she did tui-na or finger needling on me, I found it quite painful, in different places, and she pointed out that it was no longer painful to me in those places.

"That's good," she said. "You are healing!"

17. Spring Festival Healing

My second last session for the month-long intensive series of visits for acupuncture plus moxibustion, heat, finger needling, and fire cupping, with herbs and nutrition, went very well.

"You are getting better," the doctor said, " so now we do fire cupping first and then acupuncture, instead of the other way around."

She used a pair of cups on my back, up high, and then inserted the usual needles in my legs. turned on the heat lamps, and left me alone on the treatment table.

She told me a story I found quite funny.

First I told her that I had been pain-free since the last visit, and that included two very full days as I met a friend for lunch, one day, and we walked all over An Shung Mall in KaiFaQu, and the next day we met at KaiFaQu Station, took the train to Ikea and Metro, in downtown Dalian, walked all through the big stores, then took the train back to KaiFaQu, walked all over Five Color Town, had foot massages together, and then went out to dinner. After I got her a taxi, I walked from Jinma Lu through Five Color Town to KaiFaQu Station, took the train to JinShiTan, and then walked home, instead of taking a taxi.

"That's a lot of walking!" she said. "You are supposed to be resting because your back is still healing. Are you okay today?"

I told her I felt great.

She said, "Quite often, you might feel okay while walking a lot, like that, but it will be painful the next day."

"I'm a-okay," I assured her.

She confessed that the day after we walked through six museums in six hours, she felt sore the next day. "I had some pain in my back and my legs," she said. "But you say that you didn't."

"I felt fine," I said. "You should go for acupuncture," I teased her. "It really works well for back pain like that!"

She said that her husband, who does TCM, offered to do acupuncture on her, but she didn't want it.

"What?" I said. "Why not?!"

"I let him do it," she said, "but I told him not to use the big needles, that we use in China, on Chinese people, and you. That is too much for me. I can't take that, I can take only the small needles!"

We both laughed a lot about that.

She said, "I like to torture you, but I don't like to be tortured the same way!"

I assured her it was not torture.

She reminded me, again, about the first time she used the finger needling technique on me. "It was your second visit here and your friend enjoyed watching me torture you!" she said.

"Yes," I said. "I remember I was quite surprised by the whole thing. I thought you were very strong, because it hurt quite a bit."

"Now it doesn't hurt you," she pointed out. "But finger needling still hurts your friend quite a bit! That shows the difference it makes if you get a lot of acupuncture and finger needling and Traditional Chinese Medicine treatments. -- You have done a lot of healing work."

She went on to explain a few things about how finger needling works and why she pressed on points in my hands and my feet, when doing tui-na.

She explained again how the meridians work:

According to traditional Chinese medicine theory, our body consists of a giant web called the meridian system linking different parts together; its channels making up a comprehensive yet complex body map that supplies qi (vital energy) to every part of the body, assists the distribution of blood and body fluids, maintains the balance between yin and yang elements, and protects the body against disease. Along these channels, acupoints are the sites through which the qi of the organs and meridians is transported to the body surface. It is generally believed that diseases can be treated when the affected meridians or the affected organs are cleared. Acupuncturists work on these points to regulate corresponding organs or meridians so that the body can return to a state of balance and health.

The meridian system is made up by a series of channels, which are sequential to each other in the circulation of qi (vital energy). In the system, the twelve regular meridians form the major structure. They branch out to enter the chest, abdomen, and head to connect the internal organs. A total of fifteen external collaterals run along the limbs and on the trunk. There are also twelve small collaterals for controlling the muscles and tendons, and smaller collaterals disturbed on the skin surface, and the eight extra meridians to enhance the communications and functions within the system. They work closely with each other, with a dysfunction in one usually affecting another. In Chinese medicine, to be knowledgeable about the meridian system is as important as anatomy and physiology in Western medicine.

The twelve regular meridians make up the main part of the meridian system. They are distributed symmetrically on both sides of the body and are paired with their corresponding internal organs. TCM groups the meridians under arm, leg, yin and yang.

She said, "The yin meridians start in your internal organs and run from your chest, along the front of the arms, to the hands. The yang meridians start at the hands, run along the backs of the arms, and go to your head.

"Yin meridians start at the head, run down the trunk of your body, at the front, and then the front, sides, and back of your legs to the foot. And the yin meridians start at the foot, go up the inner side of the leg, cross the front of your body, at the abdomen and chest, and go to your head.

"They are all connect," she said. "So, when I do finger needling at different spots on your hands and feet, I stimulate all those meridians."

"And if it's painful," I said, "that means there's a dam in the river, or the meridians are blocked. So when you massage or finger needle those tender spots, and the pain goes away, that means the the block goes away and the dam breaks."

She nodded her head.

"So, you do acupuncture on the Bladder Meridian points and finger needling for the yin and yang meridians that go from head to toe," I said.

Again she nodded her head.

I looked up the Bladder Meridian Points and found an illustration that clearly showed where the doctor inserted the acupuncture pins in my legs.

She used GB (short for Gall Bladder) points 30 through 38.

The most sensitive, or tender, for me, was GB 30, the Acupuncture Point called Huan Tiao, in Chinese, which is called Jumping Round, in English.   The location of GB 30 is on the hips and it works for sciatica as it works on pain and numbness in the lower back buttocks, and lower limbs.

Yin Yang House, online, has great graphics illustrating the meridians and acupuncture points.

The doctor also talked about one area of difference between Western medicine and Traditional Chinese Medicine. She said, Western medicine puts people in categories but Traditional Chinese Medicine sees everybody as being quite different. In the West, they want to treat big groups of people with the same drug or surgical treatment but here we never give any two patients the same treatment.

"Hmmm," I said. "It's like that in education, as well .... Experts and administrators try to categorize students, put them in streams or levels, but teachers know it's best to get to know each student, find out where they are, and help them move along to the next stage or level."

"You must be a very good teacher!" she said. "That's the way Confucius did it!"

She talked about the Monkey King after that. The Monkey was the main character in The Journey To The West, the famous Chinese novel about Taoism and Buddhism in China and India in the Tang Dynasty era, about 1500 years ago. She laughed when she talked about the Monkey as he was a comical figure, as well as magical and a mouthpiece for both Buddhist and Taoist beliefs.

"You know, he was born from a rock," she said.

Wu Kong was a unique magical creature who was born from a rare stone egg that had soaked the power of celestial bodies for thousands of years.

I had forgotten that part of the story, I told her, and I thanked her for reminding me. The Zen master I worked with in the Zen Forest, in the Far East of Ontario, in Canada, sometimes talked about that character but he never explained what he was talking about as he assumed that I knew the story well.

Sometimes he said he thought I had been hatched from a stone egg, and I never knew what he was talking about, but suddenly a lot of things he said made a lot more sense.

I told the doctor that, and she laughed.

I told her there was a new movie, now at the An Shung movie theatre, in KaiFaQu, called Monkey King 3D and it is based on selected chapters of Wu Cheng'en's classical novel Journey to the West. It tells the story of how the Monkey King rebels against the Jade Emperor of Heaven.

According to Wikipedia, the film had the highest-grossing opening day in China with RMB121 million (US$20.0 million), surpassing Iron Man 3. It grossed RMB389.97 million (US$64.35 million) in the first four days.

It features Chow Yun-fat, from Crouching Tiger, Hidden Dragon, as The Jade Emperor, and Zang Zhilin, Miss Universe 2007, as Nüwa, a goddess in ancient Chinese mythology best known for creating mankind and repairing the wall of heaven.

"Oh, really?" the doctor said.

"It's a 3-D action film," I told her.

And that killed THAT conversation!

Jet Li was going to play the part of the Monkey King but the role went to Donnie Yen, who played against Jackie Chan in Shanghai Knights.

Donnie Yen will play the lead role in the Crouching Tiger Hidden Dragon sequel, called Crouching Tiger Hidden Dragon II – The Green Destiny.

Before I left, the doctor gave me the herbal medicine for round two, which would take me from mid-February to the end of March, she said, and we discussed the up-coming months.

She said she knew I would be busy, back at work, and would not be able to make it to the clinic for appointments every other day.

I told her that I wanted to come on Sundays for the next month as I had dentist appointments in the clinic, thanks to her, and that I might be able to make it for a mid-week appointment, now that I knew a good short-cut to get there and back.

"Good, good," she said.

We also talked about Chinese culture, or customs, a bit, as I walked out.

I learned that it is good form to give someone something using both hands. For instance, if a teacher gives a student a piece of paper using both hands, holding the paper in both hands instead of one, it is appreciated much more than if it is handed out in the quick, one-handed, or off-handed manner used in Canada and the rest of the West. However, while I was at a street food station in downtown Dalian, near Ikea, a Chinese guy and the food vendor both teased me about handing my money to her, when I paid for my jian bin, as I used both hands.

The guy spoke a few words in English and he wanted to demonstrate to me the way he handed over money for food. He used one hand. The woman he handed it to nodded her head and they shared a little laugh about the situation.

The doctor laughed, too, and instead of giving me an explanation she changed the subject to names. "When I talk to my patients from Canada," she said, "she I use their first names or their last names?"

I had to stop and think about that for a moment. Some stuffy people prefer to be addressed by others using their title and last name. Americans are more likely to use their first names than Canadians.

Earlier, we had talked about masks. I asked the doctor what she thought about people wearing masks.

I was referring to the face masks worn by many people in China. She interpreted my use of the word 'masks' differently and talked about how we use our faces as masks.

She said that when she was in California, she found it difficult to get used to the way Americans used their faces when talking to her. She said they smiled briefly, at the start of a conversation, or when introduced, but then stopped smiling.

I told her that Canadians smile a lot and Americans make fun of smiling Canucks.

She said that Chinese people smile a lot, while talking to someone, so she found it odd to talk to Americans who smiled briefly and then stopped smiling.

I told her that you can learn something about someone by turning around after you finish a conversation and walk away as the person may stop smiling and their mask may reveal their true emotion.

She said that in her experience in California you did not have to wait until the end of the conversation or walk away and then turn around because their facial expression revealed what they were thinking while they were talking to you.

She asked me what I wanted her to call me, and I said, "Marty", but then I told her I could not call her Sofia, although other Canadians did, as I thought of her, respectfully, as Dr. Wang.

She laughed and said, "Okay, Mr. Avery, that's very good!"

So, indirectly, she answered my question about handing someone something with two hands. In China, it's not necessary, but most people appreciate it and respond well to the gesture. When it comes from a Canadian, or any Westerner, it is a sign that the person has a few clues about Chinese customs and is interested in Chinese culture, so many people will be appreciative.

It made me think about the differences between East and West. If there are such differences in the use of names and facial masks, which are relatively simple things, it says something about the differences in more complex things, such as medicine and healing.

I went to the drug store and bought some face masks, which some people wear in Dalian, for a variety of reasons, and I contemplated the connections between the meridian system with acupuncture points used in Traditional Chinese Medicine and the chakras of the Ayurvedic system of India which I learned about while becoming a Reikimaster and how both systems were foreign to medicine in the West but how more and more people in Canada were learning about chakras and meridians and developing an appreciation for integrative health care, blending the best of the East and the West.

18. Not The End

After a month of TCM, with acupuncture every other day, plus moxibustion, herbs, finger needling, heat lamps, and discussion of the theory and philosophy behind it all, while getting good nutrition as well as rest and exercise, during Spring Festival in China, I felt like a new man.

Before Spring Festival, I had a few appointments, while I was busy at work. During Spring Festival, we had a five week break from work. Most of China celebrates the holiday that way. My doctor kept working, but she didn't have as many clients, since most people in China travel during the holiday. The people I work with flew south to the beaches of Bali and other countries with warmer climates.

Instead of going to Australia or someplace like that, I had over a dozen acupuncture treatments, with the other parts of TCM that go with it, and by the end of the holiday I felt as though I was flying.

I went back to work with a spring in my step and without a worry about back pain or referred pain or any pain in my brain. I walked back and forth between buildings at work, going from one campus to another, up and down a lot of stairs, and forgot about my feet, legs, butt, and back, and the fact that I had been distracted by them for well over a year.

Whenever I realized that, I felt very happy and thankful. I was full of that attitude of gratitude. And so, I want to thank Dr. Wang, once again.

Her work did create one problem: My clothes no longer fit me. I had to have a new suit made, for work.

That's a problem I was happy to deal with!

I discovered that I had dropped a size and it was easier to find clothes that fit me when I went shopping in Kaifaqu and downtown Dalian. Those stores had no clothes that fit me before that.

I had more energy, too, and felt more creative.

I've always been able to write quickly, but I was writing faster than ever.

After spending several years at a Zen retreat in the Far East of Ontario, in Canada, I wrote twice as fast as before. After several weeks of TCM in China, I was writing twice as fast again.

After my time in the Zen Forest, where I practiced meditation with massage and energy exercises, I found I could write a book a month. After TCM, I found I could write a book a week. During the Spring Festival, I wrote this book, two short novels, and two books of poetry.

And I felt younger.

I shake my head in amazement when I remember that a Canadian doctor told me I was going die and a TCM doctor in China assured me that I had very good jing-chi-shen and would likely live a long time!

Review

Acupuncture is one tool used to restore the flow of chi (qi), by inserting needles into the acupuncture points (located on the meridians). These insertions are said to clear any residing blockages, or dams, thus freeing the river to better feed the body in its entirety.

Acupuncture points employs penetration of the skin by thin, solid, metallic needles, which are manipulated manually or by electrical stimulation. "qi", that circulates through twelve invisible energy lines known as meridians on the body.

Each meridian is associated with a different organ system. An imbalance in the flow of qi throughout a meridian is how disease begins.

Acupuncturists insert needles into specified points along meridian lines to influence the restore balance to the flow of qi. There are over 1,000 acupuncture points on the body. Typically, the acupuncturist will use 6-12 needles during the treatment. The number of needles used does not correspond with the intensity of the treatment, rather it is the precise placement of the needles that is important.

Upon insertion, the client may feel a slight sting or prick. Once the needle is inserted, there should be no pain. You should feel comfortable during the treatment. If you experience pain, numbness, or discomfort, notify the acupuncturist immediately. Treatment length varies from seconds to longer than one hour. The typical length is about 20 to 30 minutes. The acupuncturist may use the following techniques during the treatment:

- Moxibustion- heating of acupuncture needles with dried herb sticks to activate and warm the acupuncture point. Also known as "moxa".

- Fire cupping - the application of glass cups to create a suction on the skin. This is to relieve stagnation of qi and blood, e.g. in sports injury.

- Herbal medicine - Chinese herbs may be given in the form of teas, pills, and capsules to supplement acupuncture treatment.

- Finger needling - Intense massage of specific points to let the river of chi flow through your body.

Moxibustion is a therapy that involves burning herbs and applying the resulting heat to specific points on the body. administered in conjunction with acupuncture.

The heat generated during moxibustion helps increase the flow of vital energy (also known as "qi" or "chi") throughout the body via "meridians".

There are two main types of moxibustion: direct and indirect. The technique most commonly used today, indirect moxibustion, often involves burning moxa (a substance created from dried leaves of the herbs mugwort or wormwood) on top of the acupuncture needle. In some cases, however, practitioners may set the burning moxa over a layer of garlic, ginger, or salt placed on the patient's skin. Other techniques include applying heat to acupuncture points from an electrical source, as well as holding the burning moxa above the skin for several minutes.

In direct moxibustion, the burning moxa is placed directly on the skin.

Twelve Major Meridians

The twelve major meridians correspond to specific human organs: kidneys, liver, spleen, hearth, lungs, pericardium, bladder, gall bladder, stomach, small and large intestines, and the triple burner (body temperature regulator). Yin meridians flow upwards. Yang meridians flow downwards. Pathways corresponding to the Yang organ is often used to treat disorders of its related Yin organ.

Tui-na or tuina is a form of massage used with acupuncture, moxibustion, fire cupping, herbs, and energy exercises. The practitioner may brush, knead, roll/press, and rub the areas between each of the joints, known as the eight gates, to attempt to open the body's chi and get the energy moving in the meridians and the muscles.   In ancient China, medical therapy was often classified as either "external" or "internal" treatment. Tui-na was considered to be one of the external methods, thought to be especially suitable for use on the elderly population and on infants. Today, Tui-na is subdivided into specialized treatment for infants, adults, orthopedics, traumatology, cosmetology, rehabilitation, sports medicine, etc.

The Meridian System

In addition to chi (qi), Traditional Chinese Medicine recognizes a subtle energy system by which chi (qi) is circulated through the body. This transportation system is referred to as the channels or meridians. There are twelve main meridians in the body, six yin and six yang, and each relates to one of the organs.

To better visualize the concept of chi (qi), and the meridians, think of the meridians as a river-bed, over which water flows and irrigates the land; feeding, nourishing and sustaining the substance through which it flows. (In Western medicine, the concept would be likened to the blood flowing through the circulatory system.) If a dam were placed at any point along the river, the nourishing effect that the water had on the whole river would stop at the point the dam was placed.

The same is true in relation to chi (qi) and the meridians. When chi (qi) becomes blocked, the rest of the body that was being nourished by the continuous flow, now suffers. Illness and disease can result if the flow is not restored.

About the author ....

Canadian author and educator Martin Avery, now living in China, wrote 100 books in Canada and is now working on 100 books in China called The Great Wall Of China Books Series. This is GWOCB #11.

Martin Avery, B.A., B.Ed.,. M.F.A., A.Q., BCTC, ATC, OTC, is a Reikimaster, Zen meditation instructor, certified Qigong leader, Quantum Touch practitioner, with a Diploma in Energy Healing.

He comes from Norman Bethune's hometown, has lived and worked across Canada, studied at universities in Canada, the U.K., and the U.S.A., and spent several years working at a Zen retreat called The Zen Forest. He now lives in JinShiTan, Dalian, China.

www.ingramcontent.com/pod-product-compliance
Lightning Source LLC
Chambersburg PA
CBHW022124170526
45157CB00004B/1745